THE ATOM BOMB

by David Killingray

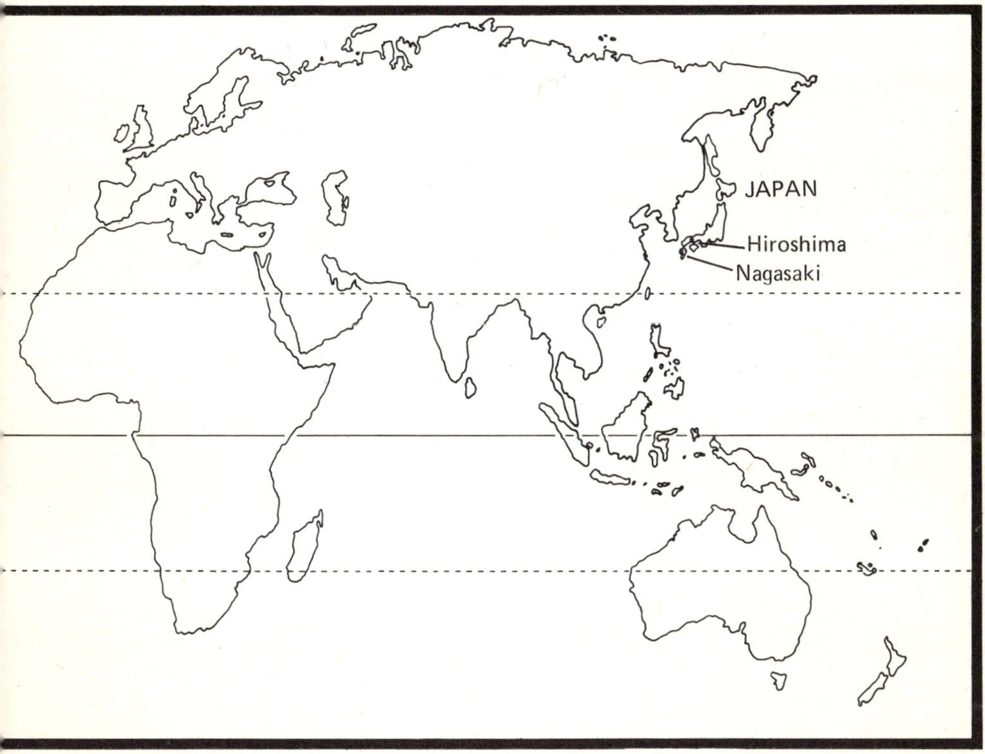

JAPAN

Hiroshima
Nagasaki

HARRAP LONDON

On the morning of 6th August 1945 an American Super-Fortress bomber flying high over the Japanese city of Hiroshima dropped the first atomic bomb. The bomb descended by parachute to a height of 600 metres, then exploded. Three days later another atomic bomb was dropped on the city of Nagasaki. Over 120,000 people were killed in the two explosions. (D1)* A few days later the Japanese surrendered. The war with the United States and China was over. *(The Two World Wars)***

A NEW AND TERRIBLE WEAPON

We expect soldiers to be killed and injured in wars. However, in nearly all wars civilians also suffer because of the fighting. In some wars civilians have been killed on purpose. During the Second World War the Nazis organized the murder of millions of Jews and Slavs who they said were inferior. *(Hitler's Reich)*

Wars in the twentieth century have involved many more civilians than previous wars. Modern warfare depends upon the produc-tion of guns, aircraft and tanks. Factory workers have become as vital as soldiers fighting at the front line. Therefore, to win a war a country has not only to win battles but also to destroy the industries of its enemy. Bombs and aircraft developed in this country have made this possible. Many people died in air raids in the Second World War. By 1944 the United States and British airforces were dropping hundreds of thousands of tons of bombs on German and Japanese cities. Three raids in two days on the German city of Dresden killed 135,000 people, and one raid on Tokyo killed more Japanese than the atomic bomb on Nagasaki.

Gas and germ warfare are other terrible ways of killing people developed in the twentieth century. These have rarely been used because they cannot easily be controlled.

Compared to previous means of destruction the atom bomb was a new and completely different type of weapon. President Truman of the United States called it a 'weapon that would revolutionize war and alter the course of history and civilization'. Ordinary bombs contain chemicals which react together to cause an explosion. Gunpowder works in a similar way when set on fire. The biggest 'blockbuster' bombs made in the Second World War contained ten tonnes of a chemical high explosive called tri-nitro-toluene, or TNT. A bomb this size could destroy a row of houses.

Atom bombs are different from ordinary bombs in several ways. Their much greater explosive power is caused by a nuclear reaction. The bomb dropped on Hiroshima had an explosive power equivalent to 20,000 tons of TNT. When ordinary bombs explode their blast and heat destroy and kill things nearby. Atomic explosions give off not only a much more devastating blast and heat but also radioactive gamma rays which can kill and injure people beyond the immediate effects of the explosion. Gamma rays damage the cell structure of the human body. A low dose of radiation may cause sickness and internal bleeding; a heavy dose can

2   *The reference (D) indicates the numbered documents at the end of this book.
   **Titles in brackets refer to other booklets in the Programme.

Atomic explosion Hiroshima: the mushroom cloud rose over seven thousand metres into the clear summer sky

Atomic bomb of the same type as 'Little Boy' which was dropped on Hiroshima. 'Little Boy' was the size of a large ordinary bomb. It was 2.04 metres long and weighed 4080 kilograms

cause cancers, which may not appear for many years, and can prevent men and women from having children. All the dust and debris in the mushroom cloud which rises above a nuclear explosion is poisoned by radiation. This 'fall-out' descends over a large area. Where it falls depends on the condition of the wind and the weather.

Since 1945 the manufacture, use, and control of nuclear weapons has been a major problem of world politics. Countries other than the United States produced atomic bombs and the costly race to develop weapons of even greater destructive power still continues. Today a *single* American or Russian bomber aircraft can carry more explosive power than all the bombs, including the two atom bombs, dropped on Japan during the Second World War. Only countries

with advanced industrial technologies are able to produce nuclear weapons. Manufacture requires large numbers of research scientists, highly skilled technicians, and huge sums of money from the government. With their large industries and many nuclear weapons, Russia and the United States are the leading nuclear states of the world. They are usually referred to as 'Super-Powers'.

THE DEVELOPMENT OF THE ATOMIC BOMB

Greek thinkers five hundred years before the birth of Christ put forward the idea that all matter consisted of minute atoms that could not be seen. From the eighteenth century onwards scientists in Europe studied atoms and tried to find out more about them. *(The Scientific Revolution)*

In the early twentieth century several important discoveries were made. French physicists discovered that uranium, a hard white metal, and another element called radium, gave off a kind of energy called radioactivity. *(Madame Curie and Einstein)* Radioactivity can be very harmful but it can be used in medicine and industry. Albert Einstein, a brilliant Swiss mathematician, put forward the new theory that vast amounts of energy were locked up in matter. If the structure of the atom could be understood it might be possible to release that energy.

For many centuries scientists had believed that the atom was the smallest part of matter and could not be split in any way. Between 1910 and 1940 scientists in industrial countries such as Germany, France, Britain, Denmark, Italy and the United States made discoveries about how the atom was made up. It was discovered that the atom consisted of a central core known as the nucleus surrounded by electrons. (D2, 3) The nucleus had two types of particles, protons and neutrons, held together by strong invisible forces known as nuclear forces. The discovery of the neutron by a British scientist in 1932 was very important. (D4) Neutrons are non-electric charges and can be used to split the nucleus of the atom. This process is known as nuclear fission.

In 1939 two German scientists, who had fled from Hitler's regime to Scandinavia, conducted experiments using neutrons to split the

Part of Ernest Rutherford's laboratory at the Cavendish Laboratory in Cambridge

uranium atom. (D5) The experiments showed that if two fragments of the split uranium nucleus flew apart at great speed they would create other neutrons which in turn would split other nuclei and generate enormous energy in a fraction of a second. This is called a chain reaction. If all the nuclei in a pound of uranium were allowed to split at the same time the energy given off would be far greater than anything ever produced by man. In fact it would be a huge explosion. (D6) However, there was a lot of difference between working out in theory how a nuclear explosion could happen and actually producing an atomic bomb.

Several major problems had to be overcome. It was vital for scientists to know how to control the number and direction of movement of neutrons. They also had to work out how to control a chain reaction or, in other words, the power released by nuclear fission. In addition adequate supplies of fissile material had to be collected. The most important was uranium which was then found mainly in Canada and the Belgian Congo, and plutonium, a recently discovered radioactive element which was man-made. These problems were solved during the Second World War when the United States developed the atomic bomb.

THE MANHATTAN PROJECT

In the summer of 1939 war broke out in Europe. Within a year the German and Italian armies controlled a large part of the continent. *(The Two World Wars)* Many scientists, some of whom were refugees from Germany and Italy, feared that Germany would develop an atomic bomb. With such a powerful weapon she would be able to win the war and become the strongest military state in the world. The scientists urged the British and American governments to speed up nuclear research and be the first to make an atom bomb. (D7) Both governments agreed to cooperate although the United States did not enter the war until late 1941.

A military atomic research project was set up in the United States. It was top secret and code-named 'The Manhattan Project'. Millions of dollars were spent on building the research stations which soon employed thousands of scientists, technicians and engineers.

Germany did not make an atomic bomb during the war. Although many of her outstanding physicists had fled abroad Germany did have a nuclear research programme. Its work was severely disrupted when essential supplies were blown up by partisans in Norway in 1942. From then on Germany concentrated her military research on producing unmanned rocket bombs. By 1945 she was a long way from manufacturing nuclear weapons.

The Manhattan Project was under the over-all control of an American general. An American physicist, J. Robert Oppenheimer, headed the research team which included some of the world's most brilliant physicists. Under the direction of Enrico Fermi, an Italian refugee, the first nuclear reactor was built in Chicago in 1942. This was an important step forward. In the reactor nuclear fission, or the production of energy from the atom, was continually taking place. A controlled chain reaction had been achieved. These successful experiments proved that it was possible to make an atom bomb and also use atomic power for peaceful means. The nuclear age had begun.

By different methods enough uranium and plutonium for use in several bombs was produced. Engineers designed the bomb and worked out how it could be detonated by charges of TNT. By the middle of 1945 the experiments were completed. (D8) On 16th July, at a site in the deserts of New Mexico, an atomic device fixed in a steel tower was exploded. (D9)

General Leslie Groves and Professor J. Robert Oppenheimer, two of the leaders of the Manhattan Project, at the site of the first atomic tests held in New Mexico. They are looking at the remains of the tower from which the test bomb was exploded

THE DECISION TO DROP THE BOMB

The Second World War was really two separate wars: one in Europe from 1939 to 1945 with Germany fighting Britain, France, Russia, and the United States; and the other in South-east Asia with Japan fighting China, Britain and the United States. Russia declared war on Japan two days after the atomic bomb was dropped on Hiroshima.

A few months before the atomic bomb test in New Mexico, Germany was defeated and occupied. Although Russia and the Western Powers — Britain, France, and the United States — had fought against Germany together they had become increasingly suspicious of one another. *(The Cold War)* By the middle of 1945 the Japanese were slowly being defeated and driven back to their home islands. Most of their navy and air force was destroyed and Japanese cities were being heavily bombed by the Americans. The Western allies and China demanded Japan's unconditional surrender. Japan refused. Most of her leaders knew that their country faced defeat but they would only accept peace on their own terms. One of their main conditions was that the Emperor should be allowed to remain as the hereditary ruler of Japan. Their attempts to get peace on these

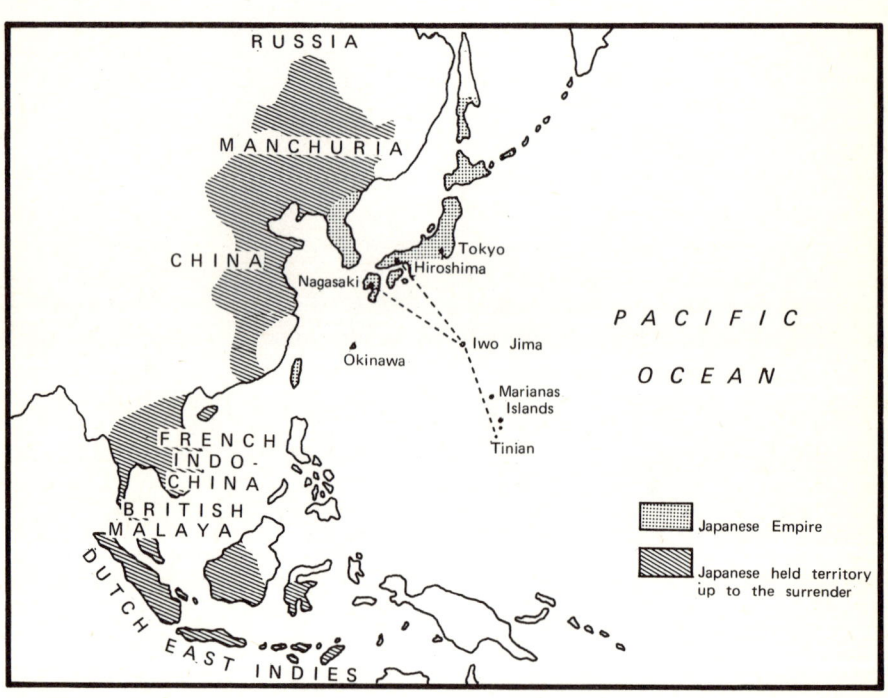

A B.29 bomber named Enola Gay *dropped the atomic bomb on Hiroshima. It took off from Tinian in the Marianas Islands and flew the 2400 km to its target via Iwo Jima*

terms, which might have led to an earlier surrender, failed. The Allies insisted on complete surrender without any terms. Russia had promised her help against Japan and she now made ready to attack the Japanese army occupying Manchuria.

The Japanese army prepared to defend their homeland. In severe battles in the Pacific islands the Japanese showed that they were prepared to fight to the death rather than surrender. On the islands of Iwo Jima and Okinawa, for example, thousands of US soldiers were killed by Japanese who fought with suicidal bravery. It was clear to American leaders that an invasion of mainland Japan would cost many lives. They argued that the atomic bomb which had been developed for use against Germany should be used to end the war with Japan quickly. (D10) The British agreed.

Some of the scientists who had helped develop the A-bomb were horrified at the idea of it being dropped on civilians and tried to persuade the American Government to demonstrate to the Japanese the power of the new weapon by using it where it would do no harm. (D11) Others said the decision should be left to the military leaders and, as the United States had only two bombs ready for use, they should not be wasted. (D12) Recently a few historians have argued that Japan would have surrendered without the atom bomb being dropped. They have also said that the United States used the bombs mainly to show Russia how powerful she was.

HIROSHIMA AND NAGASAKI

The new president of the United States, Harry Truman, made the final decision to use the atomic bombs against Japan. (D13) Two industrial cities in southern Japan were the chosen targets: Hiroshima, a military centre, with 320,000 people, and Nagasaki which had a population of 260,000.

The people of Hiroshima hurrying to work just after eight o'clock on the morning of August 6th heard the familiar sounds of the air raid sirens. Some, looking up into the clear sky, saw a long aircraft and thought it was a reconnaissance plane. Such planes occasionally flew over, but Hiroshima had hardly been bombed.

The atomic bomb exploded with a brilliant flash over the north-west centre of the city at exactly 8.16 a.m. (D14) At the heart of the explosion the temperature reached one hundred million degrees, melting metal and stone. Within five hundred metres of the explosion everything and everybody was obliterated. (D15) One thousand metres from the explosion perhaps one in every ten people, those with some form of protection, survived the searing heat and the five hundred mile-an-hour blast that tore down all in its path. At two thousand metres eight out of every ten people were killed and beyond that many others died. (D16, 17) Altogether over 40,000 people were killed outright. Within a radius of three thousand metres the blast and heat created fiercely burning 'firestorms' that took days to put out. People trapped in

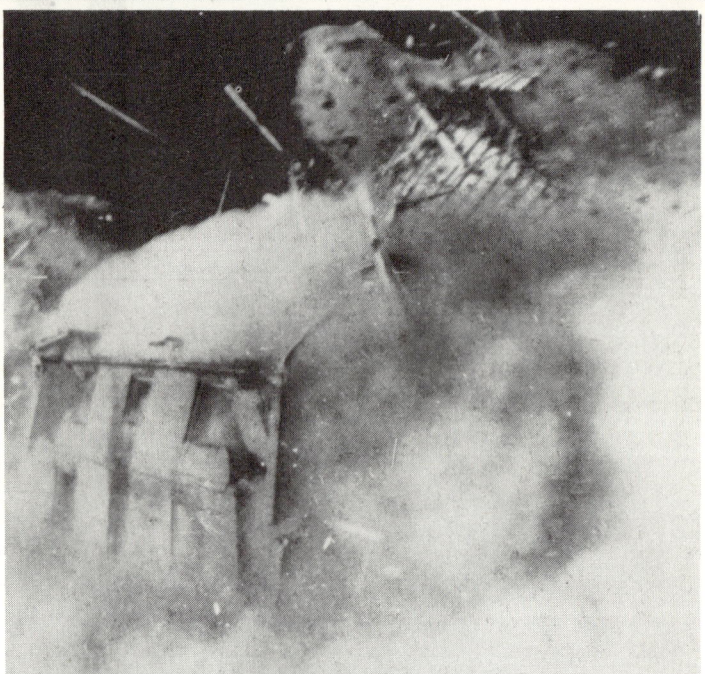

U. S. tests in the Nevada Desert in 1955 to show the effects of nuclear blasts on buildings. This two storey brick house was built 1.5 kms from the explosion. In the first picture, as the bomb is exploded, the great heat sets the house on fire. A fraction of a second later the house is hit by the blast and completely disintegrates

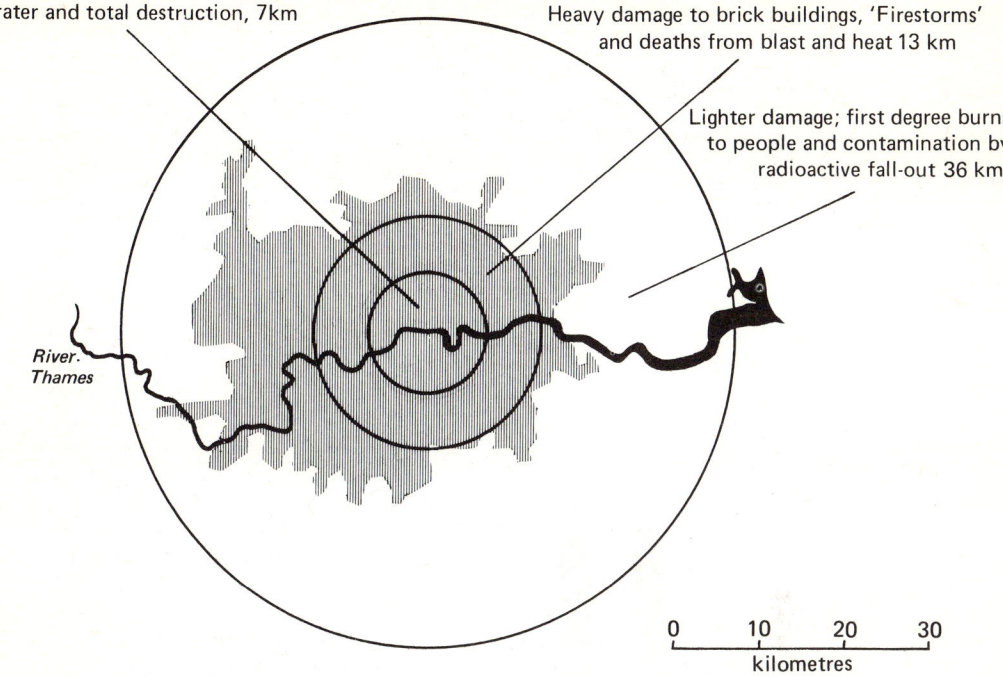

Crater and total destruction, 7km

Heavy damage to brick buildings, 'Firestorms' and deaths from blast and heat 13 km

Lighter damage; first degree burns to people and contamination by radioactive fall-out 36 km

River. Thames

0   10   20   30
kilometres

The destructive power of the hydrogen bomb. If a 10 megaton bomb exploded over central London it would kill every unprotected person within a radius of 7 km. Lethal radioactive 'fall-out' would be spread over 2500 square km. It has been estimated that it could take up to 100 years for the radio-active level to fall back to normal

the ruins of their houses were burnt to death. Streams of numbed and terrified men, women, and children fled into the surrounding countryside. The badly injured lay unattended in the charred rubble that once had been offices, houses, and streets. (D18)

About six square kilometres of the commercial and residential parts of the city were completely destroyed. Factories on the city edge were scarcely touched. Above Hiroshima a towering dark mushroom cloud rose into the sky. Radioactive 'fall-out' was scattered over a large area. Today some Japanese still suffer from the effects of that radiation. (D19)

On 9th August the Americans dropped a more powerful atom bomb on Nagasaki. Over 75,000 people died in the devastated city. The Japanese leaders, including the Emperor, knew that Japan must surrender immediately. Continued resistance might mean further bombs and a Russian invasion and occupation of the northern part of the country. The Russians were far less likely than the Americans to allow the Emperor to keep his throne. In spite of an attempt by some army officers to continue the war the Emperor announced Japan's surrender. (D20)

At the end of August the Americans began the occupation of Japan which was to bring great

11

Nagasaki after the atomic explosion, August 1945. American 'carpet' bombing raids on Japanese cities using high explosive and fire bombs caused similar devastation and many more deaths than the atomic bombs

social and political changes to the country. *(The Wealth of Japan)*

THE NUCLEAR AGE

Japan is the only country in the world against whom an atomic bomb has been used. In 1945 the United States alone had the A-bomb. This made her the most powerful military state in the world. Some scientists and politicians said that the knowledge of nuclear power should be used for the benefit of the whole world. Nuclear weapons were a danger to peace and there should be international control over their production and use. (D21) At the end of the Second World War the Western allies and Russia grew more hostile

to each other. Both sides formed alliances and spent large sums of money on producing weapons. *(The Cold War)* By 1949 Russia had developed and successfully exploded her own atomic bomb. Other industrial countries speeded up their nuclear research programmes although they had not the money or the skilled manpower to keep up with the Americans and Russians in the arms race. Britain had cooperated with the United States during the war to produce the first atomic bombs. In peace time they often disagreed over policy and Britain went ahead and made her own bomb in 1952. France exploded her first bomb in the Sahara Desert in 1960.

These bombs were all developed

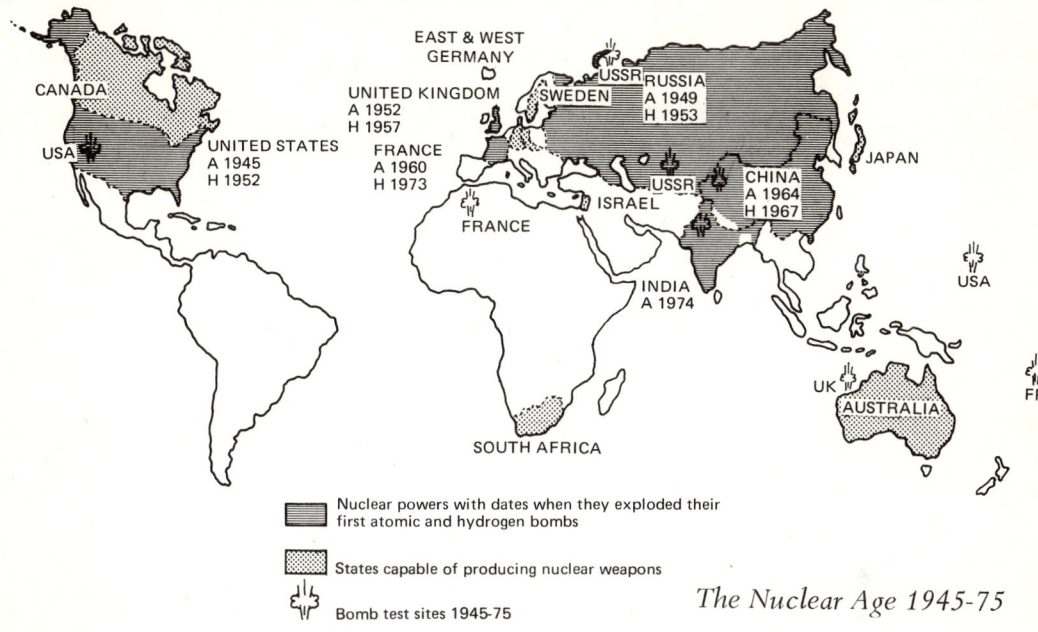

CANADA

USA
UNITED STATES
A 1945
H 1952

EAST & WEST
GERMANY

UNITED KINGDOM
A 1952
H 1957

FRANCE
A 1960
H 1973

SWEDEN

USSR

RUSSIA
A 1949
H 1953

JAPAN

FRANCE

ISRAEL

USSR

CHINA
A 1964
H 1967

USA

INDIA
A 1974

UK

AUSTRALIA

FR

SOUTH AFRICA

Nuclear powers with dates when they exploded their
first atomic and hydrogen bombs

States capable of producing nuclear weapons

Bomb test sites 1945-75

The Nuclear Age 1945-75

by advanced industrial countries who felt that it was important to have them for military and political reasons. Some countries that are able to manufacture nuclear weapons have decided not to do so. For example, Canada, Israel, Sweden, and Australia could all make atom bombs but use their nuclear research for peaceful means.

In 1964 China exploded an atom bomb. Although mainly an agricultural country she is large enough to have adequate resources with which to manufacture nuclear weapons. *(Mao Tse-Tung)* India, also a large developing country, exploded an underground nuclear device in 1974.

THE HYDROGEN BOMB

Nuclear research led to the production of the more powerful hydrogen bomb. Its explosive power was caused by two hydrogen nuclei coming together, or fusing. In the diagram of the bomb on page 14 you can see that it is detonated by chemical explosives. Hydrogen bombs can only be triggered-off by enormous heat, the kind of heat that is created by an atomic explosion. So, hydrogen bombs have atomic trigger devices built into them. These combined weapons dependent upon great heat are called 'thermo-nuclear' bombs.

The United States exploded the world's first hydrogen bomb in 1952. (D22) Russia followed in 1953. Since then Britain, France and China have also made H-bombs.

Fortunately none of these terrible weapons have been used in wars. However, they have been tested in empty parts of the world,

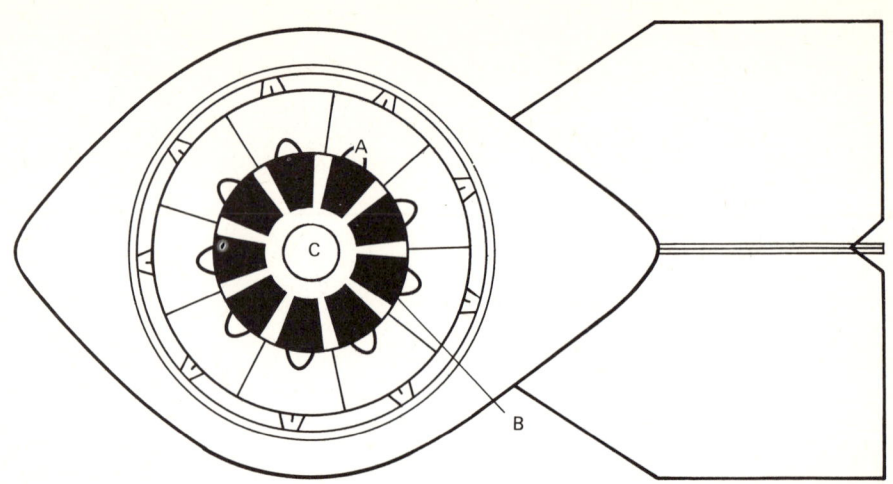

The TNT charges, 'A', explode at the same time pushing the separated wedges of plutonium, 'B', together into what is known as a 'critical mass'. This is blown into the neutron source, 'C', and explodes immediately

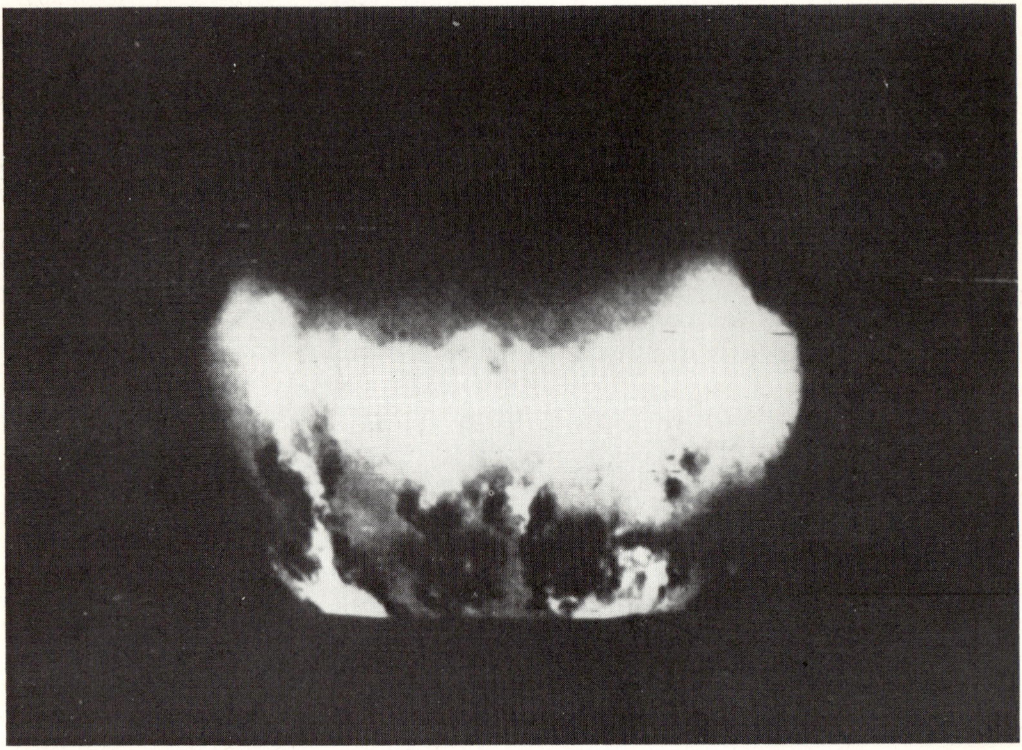

China explodes her first 'H' bomb at her testing site in Sinkiang, 1967. Although Russia in the 1960s was an ally of China she refused to give any help in developing nuclear weapons

French 'H' bombs tests in the Pacific Ocean, 1971. The nuclear device is suspended beneath the balloon

deep underground, and in the upper atmosphere. The 'fall-out' from these tests have caused a rise in radiation levels in plant and animal life. Thermo-nuclear tests are carefully carried out. But it is not always possible for those who conduct the tests to know exactly where the 'fall-out' is going to land. When the United States exploded a hydrogen bomb on Bikini Atoll in the Pacific in 1954 the 'fall-out' fell over a much larger area than was expected.

Some Japanese fishermen well outside the supposed danger zone were affected by radiation. They all became sick. Their skin turned black, their hair fell out, and one man died. (D23)

Countries have protested against nuclear tests in their areas. West African states condemned France at the United Nations for exploding bombs in the Sahara. More recently New Zealand and Australia have protested against French tests in the Pacific.

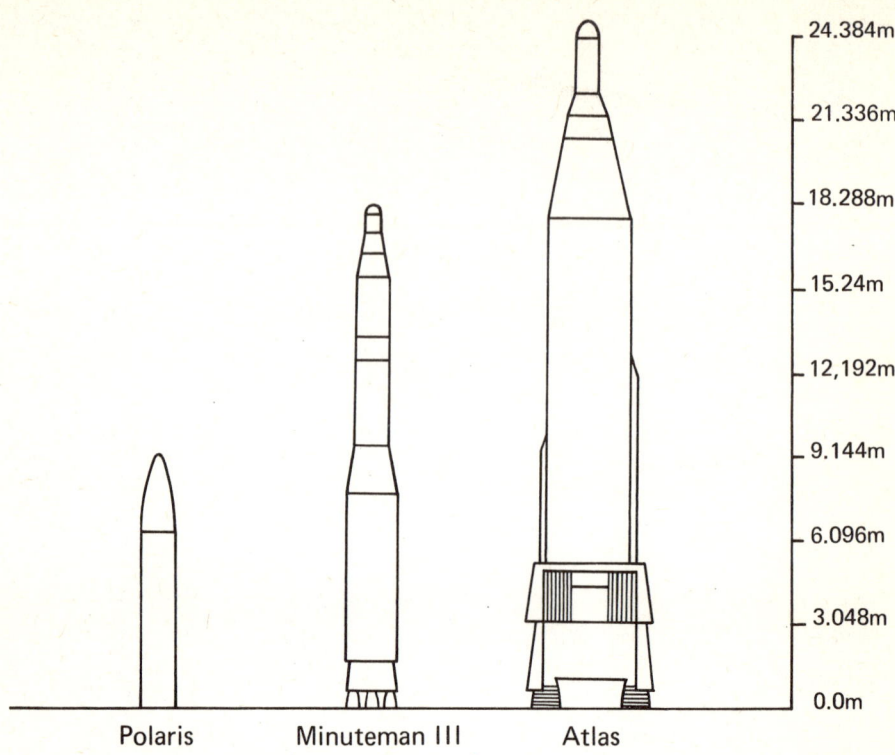

	24.384m
	21.336m
	18.288m
	15.24m
	12,192m
	9.144m
	6.096m
	3.048m
	0.0m

Polaris   Minuteman III   Atlas

Three Intercontinental Ballistic Missiles

NUCLEAR MISSILES

A large part of the costly scientific research on nuclear weapons has gone into developing new ways of delivering bombs onto their targets, and ways of defending a country against attack.

Aircraft were used to drop the atomic bombs on Japan. This became an out-dated method of delivery. Although bombers can now fly at over 2000 miles per hour they are not as fast as short-range electronically guided rockets, or missiles, which have become part of modern defence systems. These rockets are rather like super explosive bullets which can find and destroy attacking aircraft.

Instead of aircraft, long-range missiles were developed to deliver nuclear bombs. The explosive part of the missile was known as the nuclear warhead. Inter-continental missiles (ICBM), first built by the Russians in 1957, can fly thousands of miles at speeds greater than 20,000 miles per hour and be guided onto their targets. Unlike manned aircraft they do not have to return. Research in rocket technology by Russia and the United States was linked to their space programmes.

The world powers have thousands of missiles with nuclear warheads that can be launched from trucks, nuclear-powered submarines under the sea, and sites

A Poseidon missile being fired from a United States nuclear-powered submarine

deep in the ground. In the past if one country went to war with another it might take days, weeks, or even months before heavy military blows were struck. Today the great nuclear powers — Russia, China, and the United States — all have armed missiles aimed at each other. At the press of a button terrible destruction can be unleashed in minutes.

Both the United States and Russia seek to deter the other from starting a major war by having

17

A United States B.52 bomber launching a 'Hound Dog' intercontinental missile. 'Hound Dog' is about 13.5 metres long and has a nuclear warhead

approximately equal power in nuclear weapons. Thus, in a way, world 'peace' is kept by nuclear deterrent. This means that the super-powers try to keep up with each other in developing more powerful nuclear weapons, faster rockets, and better defence systems. Both countries have complex radar warning systems, rockets designed to intercept attacking missiles (anti-missile missiles), and plans to carry on the government from underground shelters in event of war. They have also produced smaller nuclear weapons, known as 'tactical' weapons, which have a low explosive power. Weapons of this sort could be used in a limited non-nuclear war. The great

powers agreed that they should not be used. In August 1945 the United States had only two atomic bombs ready for use. In 1974 the great powers have huge stock-piles of nuclear weapons of all kinds equivalent to several thousand million tonnes of TNT.

Between the two World Wars several international disarmament conferences tried to control the manufacture and use of weapons. They achieved very little. Because nuclear weapons are so much more destructive than ordinary weapons it is less likely that a great power would seek a full-scale war to achieve its political aims. After 1950 many countries, particularly Russia and the United States, have

Intercontinental ballistic missiles in a Russian military parade through Red Square, Moscow, 1963

In the 1950s-60s well-organized groups in many countries tried to persuade governments to stop the manufacture and testing of atomic weapons. This march by the British Campaign for Nuclear Disarmament in 1962 is led by two Japanese survivors from Hiroshima

Signing the Nuclear Test Ban Treaty in the Kremlin, Moscow, 1963. Foreign ministers from left to right are: Dean Rusk (USA), Andrei Gromyko (USSR) and Lord Home (UK)

tried to control the spread, manufacture, and use of nuclear weapons. They felt that if more countries developed and tested atom bombs then there would be an increase in radioactive pollution and a greater risk of nuclear war. The United Nations has taken a leading part in disarmament talks. Since China produced atomic weapons Russia and the United States have cooperated more closely on disarmament. In 1963, along with sixteen other countries, they agreed to ban nuclear tests in the atmosphere, the sea, and in outer space. (D24, 25) A further treaty to prevent the spread of nuclear weapons was signed in 1968. China and France, who had not fully tested their bombs, refused to sign either treaty.

THE PEACEFUL USES OF NUCLEAR ENERGY

Thirty years ago scientists hoped that nuclear energy would soon provide the world with plenty of cheap power. But progress has been slower than expected. Although one pound of uranium contains the same amount of energy as several hundred tonnes of coal, releasing that energy is a costly and difficult scientific problem. Industrial countries would like to produce cheap and plentiful supplies of electricity from nuclear power stations. Most of their electricity is made with coal and oil. There are only certain amounts of these fuels in the ground and each year as more is used up it becomes important to

Wylfa nuclear power station, North Wales, United Kingdom. Built in the late 1960s this power station provides power for the national electricity grid

find other sources of energy. Nuclear power stations in several countries already produce some electricity. They are usually built in remote places in case of accidents which might result in radiation escaping. (D26)

The atom has many peaceful uses. Some cargo ships have nuclear engines and one day it may be possible for cars and trains to be powered in the same way. Nuclear processes have helped to develop stronger metals and plastics and controlled nuclear explosions could be used to build canals, harbours, and dams.

Two major problems of the peaceful uses of nuclear energy are pollution and the disposal of radioactive waste from power stations and industries. Such waste remains radioactive and dangerous for a very long time. It cannot just be dumped into the sea or buried underground. Nor can it be cheaply and easily made 'safe'.

But radioactivity can be put to good uses. Radioactive isotopes (a substance with an unstable atom which gives off radioactivity) are of use in medicine, engineering, chemical research, archaeology, and agriculture. For example, cancers can be controlled, strains of seeds improved, cracks in buried oil pipes located, and disease carrying pests destroyed. And there may be many other ways in which the tiny atom can help mankind. For scientists have found out only a little about the great power of nuclear force.

21

DOCUMENT 1

DROPPING THE BOMB   *ABE SPITZER — A B-29 radio operator who flew on both the Hiroshima and Nagasaki missions, describing the reactions of himself and his fellow airmen*

'The part that got me most', Ray commented, 'was when I saw that ball of fire. I guess I can tell you now, and you won't laugh, but I thought maybe the world had come to an end, and we'd caused it.'
'I can't forget the pillar of smoke,' said Buckley, 'reaching all the way up . . . to the top of the sky, and I kept wondering if it would ever stop . . . .'
'Funny,' added Pappy, 'I couldn't help thinking about the people who'd been living in those houses down there. They never knew what hit them. I knew they were Japs, of course, and you don't think much about Japs. They're your enemies and all. But I kept thinking about that, and I still do. I guess I always will. Well, a world did come to an end, although we hadn't caused it. We were under orders; we were just doing our job. And I don't think Ray meant it in exactly the way I do, but he was close enough . . . . The missions to Hiroshima and Nagasaki are nothing to be proud of at all. But they are missions to be remembered. And never repeated. Anywhere in the world. Ever again.'

DOCUMENT 2

IDENTIFYING THE ELECTRON   *J.J. THOMSON — British physicist who discovered in 1899 the first particle in the atom, which he called a 'corpuscle' and which was later named the electron*

I regard the atom as containing a large number of smaller bodies which I will call corpuscles; these corpuscles are equal to each other; the mass of the corpuscle is the mass of the negative ions in a gas at low pressure.

DOCUMENT 3

THE STRUCTURE OF THE ATOM   *NIELS BOHR — A Danish physicist describing in 1913 Professor Rutherford's theory which showed that the atom was not solid but largely empty*

Professor Rutherford has given a theory of the structure of the atom. According to this theory the atoms consist of a positively charged nucleus surrounded by a system of electrons kept together by attractive forces from the nucleus; the total negative charge of the electron is equal to the positive charge of the nucleus. Further the nucleus is assumed to be the seat of the essential part of the mass of the atom, and to have linear dimensions exceedingly small compared with the linear dimensions of the whole atom.

DOCUMENT 4

THE EXISTENCE OF THE NEUTRON   *PROFESSOR J. CHADWICK*
Describing an experiment at Cambridge University, 1932

I made further experiments to examine the proportion of radiation excited in beryllium. [These] were first examined by means of . . . a small ionisation chamber connected to a valve amplifier . . . . It is evident that we must adopt another idea about the nature of radiation . . . [that it] consists of particles of mass very nearly equal to that of the proton . . . . In order to explain the great penetrating power of the radiation we must further assume that it has no net charge . . . and consists of a proton and an electron in close combination, the 'neutron'. We must . . . suppose that the neutron is a common constituent of atomic nuclei . . . the neutron hypothesis . . . throws new light on the problem of nuclear structure.

DOCUMENT 5

NUCLEAR FISSION   *PROFESSOR OTTO FRISCH – Describing the end of his experiment to prove the fission of the uranium nucleus, 1939*

I worked until three in the morning. Many checks had to be carried out. From time to time I removed my neutron source to make sure the counter stopped ticking . . . . Once I took the uranium out of the chamber, and again the counter was silent, as I had expected. Could my batteries have run down? I put the uranium back, and the counter clicked again, once or twice a minute. It was slow and tedious work, but I assure you I was not bored! When I went to bed I was dead tired, but very happy; I had obtained clear physical evidence for the fission of the uranium nucleus.

DOCUMENT 6

THE FRISCH-PEIERLS MEMORANDUM, 1940   *Two German scientists working in Britain suggested that a 'Super-Bomb' could be constructed based on a nuclear chain reaction in uranium*

We have . . . come to the conclusion that a moderate amount of 235U [uranium] would indeed constitute an extremely efficient explosive . . . one might think of about 1 kg [of 235U] as a suitable size for the bomb . . . . The energy liberated by a 5 kg bomb would be equivalent to that of several thousand tons of dynamite . . . .
    In addition to the destructive effect of the explosion itself, the whole material of the bomb would be transformed into a highly radioactive state . . . . the radiations would be fatal to living beings even a long time after the explosion . . . . Any estimates of the effects of this radiation on human beings must be rather uncertain because it is difficult to tell what

will happen to the radioactive material after the explosion. Most of it will probably be blown into the air and carried away by the wind. This cloud of radioactive material will kill everybody within a strip estimated to be several miles long . . . . If 1% of the active material sticks to the debris in the vicinity of the explosion and if the debris is spread over an area of, say, a square mile, any person entering this area would be in serious danger, even several days after the explosion . . . . Effective protection is hardly possible . . . . The irradiation is not felt until hours later when it may be too late.

DOCUMENT 7

THE MAUD COMMITTEE   *A report to the British Government on The Use of Uranium for a Bomb, 1941*

1   The committee considers that the scheme for a uranium bomb is practicable and likely to lead to decisive results in the war.
2   It recommends that this work be continued on the highest priority and on increasing scale necessary to obtain the weapon in the shortest possible time.
3   That the present collaboration with America should be continued and extended . . . .

DOCUMENT 8

THE ATOM BOMB   *GROUP-CAPTAIN LEONARD CHESHIRE — The official British observer on the Nagasaki raid*

To me the most fantastic spectacle . . . was when Alvarez walked me across to the box and opened it. After all that I had heard and read during the past few weeks . . . I should have expected to find [the bomb] entrenched behind steel and concrete. I should have expected it to be high on a pedestal or deep in the bowels of the earth; but not just lying casually in a box in an ordinary Nissen hut. Equally, I should not have expected the man who showed it to me to open the box in the course of conversation, and after I had glanced at the contents and looked away again because it did not seem to be anything particularly interesting, to say, 'That is the atom bomb', as though it were straw or anything else that is normally found in boxes.

DOCUMENT 9

THE FIRST TEST EXPLOSION   *GENERAL THOMAS FARRELL —*
Describing the first Atom Bomb explosion in New Mexico in 1945

The whole country was lighted by a searing light with an intensity many times that of the midday sun . . . . Thirty seconds after, the explosion came, first the air blast pressing hard against the people and things, to be followed almost immediately by the strong, sustained awesome roar which warned of doomsday and made us feel that we puny things were blasphemous to dare tamper with the forces heretofore reserved to the Almighty.

DOCUMENT 10

THE DECISION TO DROP THE BOMB   *HENRY STIMSON —*
US Secretary of War in August 1945

The principal . . . objective of the United States in the summer of 1945 was the prompt and complete surrender of Japan . . . . In the middle of July 1945, the intelligence section of the War Department General Staff estimated Japanese military strength . . . at about 5,000,000 men . . . . As we understood it in July, there was a very strong possibility that the Japanese Government might determine upon resistance to the end . . . In such an event the Allies would be faced with the enormous task of destroying an armed force of five million men . . . belonging to a race which has already amply demonstrated its ability to fight literally to the death.
 The . . . plans . . . for the defeat of Japan . . . had been prepared without reliance upon the atomic bomb . . . We estimated that . . . the major fighting would not end until the latter part of 1946, at the earliest. I was informed that such operations might be expected to cost over a million casualties to American forces alone. My chief purpose was to end the war in victory with the least possible cost in the lives of the men in [our] armies . . . . The decision to use the atomic bomb was a decision that brought death to over a hundred thousand Japanese. No explanation can change that fact and I do not wish to gloss over it. But this deliberate, premeditated destruction of Hiroshima and Nagasaki put an end to the Japanese war. It stopped the fire raids, and the strangling blockade; it ended the ghastly spectre of a clash of great armies.

DOCUMENT 11

THE USE OF THE BOMB 1   *Seven atomic scientists urged the US Secretary of War not to use the Atomic Bomb against Japan, but to demonstrate its power to them*

Thus, from an 'optimistic' point of view — looking forward to an international agreement on the prevention of nuclear warfare — the military advantages and the saving of American lives achieved by the sudden use of atomic bombs against Japan may be outweighed by the ensuing loss of confidence and by a wave of horror and repulsion sweeping over the rest of the world and perhaps even dividing public opinion at home.

From this point of view, a demonstration of the new weapon might be made, before the eyes of representatives of all the United Nations, in the desert or on a barren island. The best possible atmosphere for the achievement of an international agreement could be achieved if America could say to the world, 'You see what sort of a weapon we had but did not use. We are ready to renounce its use in the future if other nations join us in this renunciation and agree to the establishment of an efficient control.'

DOCUMENT 12

THE USE OF THE BOMB 2   *Comments by a scientific panel advising the US Government about the A-Bomb in 1945. The panel included Enrico Fermi and Robert Oppenheimer*

We were asked to comment on whether the bomb should be used . . . . We didn't know beans about the military situation in Japan. We didn't know whether they could be caused to surrender by other means or whether the invasion was really inevitable . . . . We said that we didn't think that being scientists especially qualified us as to how to answer this question of how the bombs should be used or not . . . . We thought the two overriding considerations were the saving of lives in the war and the effect of our actions on the stability of our strength and the stability of the post-war world. We did say that we did not think that exploding one of these things as a firecracker over a desert was likely to be very impressive.

DOCUMENT 13

THE DECISION TO USE THE BOMB   *PRESIDENT TRUMAN*

The final decision of where and when to use the atomic bomb was up to me. Let there be no mistake about it. I regarded the bomb as a military weapon and never had any doubts that it should be used. [My] top military advisers recommended its use, and when I talked to Churchill, he unhesitatingly told me that he favoured the use of the atomic bomb if it might aid to end the war . . . .

DOCUMENT 14

HIROSHIMA   *A Japanese professor, who was 5000 metres from the centre of the explosion, describes what happened*

A blinding flash cut sharply across the sky. I threw myself onto the ground. At the same moment as the flash, the skin over my body felt a burning heat. Then there was dead silence, probably for a few seconds, and then a huge 'boom' like the rumbling of distant thunder. At the same time a violent rush of air pressed down my entire body. I raised my head and saw an enormous mass of clouds spreading and climbing rapidly into the sky where it took on the shape of a monstrous mushroom.

DOCUMENT 15

HIROSHIMA   *A description of the city the day after the bomb was dropped*

I climbed Hijiyama Hill and looked down. I saw that Hiroshima had disappeared. I was shocked by the sight. What I felt then and still feel now I just can't explain with words. Of course, I saw many dreadful scenes after that — but that experience, looking down and finding nothing left of Hiroshima — was so shocking that I simply can't express what I felt. For acres and acres the city was like a desert except for scattered piles of brick and roof tiles. I could see the suburb of Koi and a few buildings standing . . . but Hiroshima didn't exist, it just didn't exist.

DOCUMENT 16

THE EXPLOSION   *TOSHIO NAKAMURA — An eleven-year-old child writing about the explosion a year after the bomb was dropped*

The day before the bomb, I went for a swim. In the morning, I was eating peanuts. I saw a light. I was knocked to little sister's sleeping place. When we were saved I could only see as far as the tram. My mother and I started to pack our things. The neighbours were walking around burned and bleeding . . . . We went to the park. A whirlwind came. At night a gas tank burned and I saw the reflection in the river. We stayed in the park one night. Next day I went to the Taiko Bridge and met my girl friends Kikuki and Marukami. They were looking for the mothers. But Kikuki's mother was wounded and Marukami's mother, alas, was dead.

DOCUMENT 17

THE EXPLOSION   *YOKO OTA — Japanese woman writer describing her immediate reaction to the explosion*

I just could not understand why our surroundings had changed so greatly in one instant. I thought it might have been something which was nothing to do with the war, the collapse of the earth which it was said would take place at the end of the world, and which I had read about as a child.

DOCUMENT 18

THE AMERICAN VIEW   *Report in the* New York Times, *7th August 1945*

FIRST ATOMIC BOMB DROPPED ON JAPAN
MISSILE IS EQUAL TO 20,000 TONS OF TNT
TRUMAN WARNS FOE OF A 'RAIN OF RUIN' . . .
HIROSHIMA IS TARGET

Editorial: *Our Answer to Japan*

The American answer to Japan's contemptuous rejection of the Allied surrender ultimatum of July 26 has now been delivered upon Japanese soil in the shape of a new weapon of destruction which unleashes against it the forces of the universe . . . . What every military man has dreamed of . . . — the decisive 'secret weapon', the magic key to victory — has been found in America and is now ready to be hurled against our enemies.

So far only one of these bombs has been dropped . . . . That is just a sample of what is in store for Japan . . . .

DOCUMENT 19

RADIATION SICKNESS   *FUMIO NAKAMURA — An inhabitant of Hiroshima writing in a letter just before his death in 1958*

Thank you for being upset about me. I am spending very miserable days since the beginning of January. I feel as if my whole body is rotting away from my internal organs. As time goes by I feel weaker and weaker. I feel as if I am being cooked alive. My condition is becoming worse day by day and I'm unable to work . . . . To suffer with a disease which has *no cure* is like drinking poison every day and waiting for my life to end. Sometimes I feel so bad that I roll about in my bed all night. I do feel miserable!

DOCUMENT 20

JAPAN'S SURRENDER *EMPEROR HIROHITO – Broadcasting,*
15th August 1945

To Our good and loyal subjects
After pondering deeply the general trends of the world and the actual
conditions obtaining in Our Empire today, We have decided to effect a
settlement of the present situation by resorting to an extra-ordinary
measure.
We have ordered Our Government to communicate to the Governments
of the United States, Great Britain, China and the Soviet Union that
Our Empire accepts the provision of their Joint Declaration [the Potsdam
agreement] . . .
Despite . . . the gallant fighting of military and naval forces . . . the war
situation has developed not necessarily to Japan's advantage, while the
general trends of the world have all turned against her interest . . . . the
enemy has begun to employ a new and most cruel bomb, the power of
which to do damage is indeed incalculable, taking the toll of many
innocent lives. Should we continue to fight, it would not only result in
an ultimate collapse and obliteration of the Japanese nation, but also it
would lead to the total extinction of human civilization.

DOCUMENT 21

INTERNATIONAL CONTROL OF THE ATOM BOMB *HENRY
STIMSON – Retiring US Secretary of War in a memorandum to the
new president Harry Truman, September 1945*

The introduction of the atomic bomb has profoundly affected political
considerations in all sections of the globe . . . . It appears that Russian
influence has been weakened . . . and there will be a strong temptation
for the Soviet leaders to acquire this weapon in the shortest possible
time . . . . This will almost certainly cause a secret armament race of a
rather desperate character . . . . What is important to the world and
civilization is to make sure that when the Russians do get these weapons
they are willing and cooperative partners among the peace-loving nations
of the world . . . . American-Soviet relations may be perhaps irretrievably
embittered by the way in which we approach the solution of the bomb
with Russia. For if we fail to approach them now and merely continue
to negotiate with them, having this weapon rather ostentatiously on our
hip, their suspicions and their distrust of our purposes and motives will
increase . . . .

DOCUMENT 22

DEVELOPING THERMONUCLEAR WEAPONS   *Two of the nine*
scientists on the General Advisory Commission reporting to the US
Government 1950

The fact that no limits exist to the destructiveness of this weapon makes
its very existence and knowledge of its construction a danger to humanity
as a whole. It is necessarily an evil thing considered in any light. For these
reasons we believe it important for the President of the United States to
tell the American public and the world that we think it wrong on
fundamental ethical principles to initiate the development of such a
weapon.

DOCUMENT 23

THE LUCKY DRAGON   *Although 90 miles away from an H-Bomb*
test the crew of the Lucky Dragon were affected by radioactive
'fall-out'

Several tons of tuna fish, stated to be dangerously radioactive, were seized
here [Yaizu] and Osaka to-day. The fish had come from the Japanese
fishing vessel Fukuru Maru [Lucky Dragon], whose crew are suffering
from burns and scars believed to have been caused by radioactivity from
an American hydrogen bomb test explosion in mid-Pacific on Bikini Atoll
on March 1. The health authorities are searching for six members of the
crew, who have not been seen since they went ashore on Sunday. It is
feared that unless they are found they will contaminate others. Seventeen
others are in hospital suffering from burns and scars . . . . One of the [men]
is in serious condition. His face is black with burns, his ears are covered
with scabs which have been suppurating, and his hands are swollen. All the
others had burns, some of their hair had fallen out, and they displayed
other symptoms of radiation sickness.

DOCUMENT 24

NUCLEAR DISARMAMENT   *NIKITA KRUSHCHEV – The Russian*
leader in a speech to the Supreme Soviet, January 1960

It is perfectly clear to all sober-minded people that atomic and hydrogen
weapons are particularly dangerous to the countries that are densely
populated. Of course, all countries will suffer in one way or another in
the event of a new world war. We, too, shall suffer much, shall sustain
great losses. Our territory is immense and our population is less concentrated in large industrial centres than is the case in many other countries.
The west will suffer incomparably more . . . .

DOCUMENT 25

THE NUCLEAR TEST BAN TREATY   *PRESIDENT KENNEDY –*
Broadcasting to the American people 1963

I speak to you tonight in a spirit of hope. Eighteen years ago the advent
of nuclear weapons changed the course of the world . . . . Since that time,
all mankind has been struggling to escape from the darkening prospect
of mass destruction on earth . . . an age when both sides . . . possess
enough nuclear power to destroy the human race several times over . . . .
Yesterday a shaft of light cut into the darkness. Negotiations were con-
cluded [with Russia] . . . to bring the forces of nuclear destruction under
international control . . . it offers to all the world a welcome sign of hope
. . . a step away from war.

DOCUMENT 26

RADIATION LEAKAGE   *In October 1957 there was a small explosion*
at the British nuclear power station of Windscale. This is part of an
official bulletin issued by the Atomic Energy Authority (AEA)

Midnight 10th October 1957:
Since the recent mishap at Windscale extensive measurements have been
made of the radioactivity in the district. It is confirmed that the general
levels of radioactivity in the district are not dangerous. But in the case of
milk, which after a few days begins to concentrate whatever radioactivity
is present, the AEA have arranged . . . with the Milk Marketing Board for
the temporary suspension of distribution from farms near the works. The
level of radioactivity in the milk from these farms is expected to rise for
a few days and to fall rapidly thereafter.

ACKNOWLEDGMENTS

Illustrations

Associated Press Ltd pages 7, 10, 19 bottom; Cavendish Laboratory, University of Cambridge page 5; Imperial War Museum pages 3, 4, 12; Keystone Press Ltd pages 14 bottom, 15, 17, 18, 19 top, 20; United Kingdom Atomic Energy Authority page 21.

Documents

D1, *We Dropped the A-Bomb,* Merle Miller and Abe Spitzer, McIntosh and Otis Inc; D4, *The Existence of the Neutron,* J. Chadwick, The Royal Society; D5, *Working with Atoms,* Otto Frisch, Brockhampton Press Limited; D6, 7, *Britain and Atomic Energy,* Margaret Gowing, Macmillan; D8, *The Spectator, 6th August 1956,* T. Cheshire; D9, 11, 12, 22, *Brighter Than a Thousand Suns,* Robert Jungk, Victor Gollancz Ltd; D13, *Year of Decisions 1945,* Truman, Hodder and Stoughton Limited; D14, 15, 17, *Death in Life; The Survivors of Hiroshima 1968,* Robert Lifton, George Weidenfeld and Nicolson; D16, *Hiroshima 1946,* John Hersey, Hamish Hamilton; D19, *The Devil's Repertoire,* Victor Gollancz; D20, *Japan's Decision to Surrender,* J. C. Butow, Stanford University Press; D21, *On Active Service in Peace and War,* Henry L. Stimson and McGeorge Bundy, Harper & Row; D24, *Soviet Foreign Policy Since the Death of Stalin,* trans. H. Hanak, Routledge & Kegan Paul Ltd; D25, *The Politics of John F. Kennedy,* Edmund Ions, Routledge & Kegan Paul Ltd.